GUNPILOT: A VIETNAM P.O.W.

CAPTAIN WILLIAM S. REEDER JR.

BY E. DANIEL KINGSLEY CW4, US ARMY, RETIRED

Edited by
ELIZABETH KINGSLEY

Edited by
DROLLENE P BROWN

THRIVING RESULTS COACHING

GUNPILOT: A Vietnam P.O.W.

Captain William S. Reeder, Jr.

© 2025 by E. Daniel Kingsley, All rights reserved.

Published by Thriving Results Publishing

Orem, Utah

Cover Design by Steven Novak Illustrations

Editing by Drollene P. Brown and Elizabeth Kingsley

About the Author (Chapter 11) taken from https://veterantributes.org/

ISBN (trade soft): 978-1-967014-04-0

ISBN (ebook): 978-1-967014-02-6

PREFACE

I met Bill Reeder when he was the Executive Officer of 9th Combat Brigade, Air Assault (CBAA), at Fort Lewis, Washington, in 1983. I suppose I became one of Bill's greatest fans.

At that time, I was an aviation safety Warrant Officer in his subordinate cavalry unit (3/5 Cav), just pretending to be a writer. I had seen him around the brigade.

His reputation was one of those "Hell of a Man," but he looked pretty plain for a "Hell of a Man." He was lean, quiet, always seemed happy, and stoic as hell. He never complained about anything except soldiers who didn't take care of soldiers.

We weren't exactly close, but he was the Brigade Exec, and you see things. You notice things.

I was with him once on a brigade run. A brigade run for the 9th meant every man in the brigade ran three miles in formation. On this run, Bill finished limping badly but with no accompanying grimace. He had fallen back in the formation. He finished with one of the battalions rather than up front with the command staff. It was unusual.

I heard later that it was his war injuries, specifically prisoner of war (POW) injuries, that caused his limp. It was said that he had refused to drop out as a matter of principle.

I was impressed. I was an aspiring writer and wanted to write about a POW. So I asked him for an interview.

He said yes because he wanted to share all of the POW lessons he learned with other soldiers. We spoke for three hours. I recorded the interview. That was in 1983.

I never got a 'round tuit' and wrote it.

Eighteen years later, I found those audio tapes. I dug them out of the box. With his renewed permission, I have proceeded to write.

This little book is his story. It is the most detailed account of his experience ever published.

It is *not* intended to be a transcript, but rather a faithful, concise telling of all the facts. Reeder's account is the story of a real soldier. There is little fluff and less pretense in it. He takes little credit, while he gives credit by name to so many who crossed his path. He credits those who lent him their strength during the trial.

Bill Reeder's motives in letting me write about it are purely professional. It *is* intended to inform a professional soldier audience. Therefore, I've included many relevant facts that are not necessary for the enjoyment of the lay audience, with extra details of interest in footnotes.

This is not a story of years of captivity. Nor is it of years of torture. Reeder actually arrived in Hanoi after the worst of torture had been stopped. Every POW in Vietnam experienced unique circumstances. There are lengthier and certainly more tortured POW experiences.

Col Bill Reeder wanted his experiences to be used by other soldiers, for their learning and benefit in surviving enemy capture and/or torture. To that end, I have included additional POW reading-list, for the person who wants to be well rounded in the understanding of what the term POW means.

You cannot prepare for torture, but you can understand the basic honorable POW concept. You can learn to keep the faith; you can learn to take every advantage; and you can ensure that if it costs your life, the enemy pays a price.

But a soldier can survive being shot, bombed, stabbed, burned or run over. That soldier can live through disease and starvation. If that soldier is not frozen, drowned, bitten by something poisonous or eaten, then he may use the books listed here to good effect in any war.

Each of the books on the list are very valuable. All of them point at the survival motive to endure to the end, and how it can be done. Each POW story teaches something different.

This is a true story of courage in the face of death and uncertainty all around. I knew Bill Reeder. He is retired now as a full Colonel, loves life and shares it with those around him. He has earned all the richness life can bring.

E. Daniel Kingsley

OTHER RECOMMENDED READING:

**5 Years to Freedom** by James N. Rowe

**POW** by John G. Hubbell

**The Passing of the Night** by Robinson Risner

**Through the Valley: My Captivity in Vietnam** by William S Reeder, COL, Retired, US Army (Annapolis, MD: Naval Institute Press), 2016.

CODE OF CONDUCT

I

I am an American, fighting in the forces which guard my country and our way of life. I am prepared to give my life in their defense.

II

I will never surrender of my own free will. If in command I will never surrender the members of my command while they still have the means to resist.

III

If I am captured I will continue to resist by all means available. I will make every effort to escape and aid others to escape. I will accept neither parole nor special favors from the enemy.

IV

If I become a prisoner of war, I will keep faith with my fellow prisoners. I will give no information or take part in any action which might be harmful to my comrades. If I am senior, I will take command. If not, I will obey the lawful orders of those appointed over me and will back them up in every way.

V

When questioned, should I become a prisoner of war, I am required to give name, rank, service number, and date of birth. I will evade answering further questions to the utmost of my ability. I will make no oral or written statements disloyal to my country and its allies or harmful to their cause.

VI

I will never forget that I am an American, fighting for freedom, responsible for my actions, and dedicated to the principles which made my country free. I will trust in my God and in the United States of America.

1

SHOT DOWN

IT ALL STARTED IN 1965.

The Vietnam War was just cranking into full gear. Barry Goldwater had run for president on a platform that said we needed to get *out* of Vietnam. Goldwater had warned that Lyndon Johnson would take us deeper into that war. That is exactly what happened. After all was said and done, the whole country was of the "let's kick their Commie butts" attitude as politics dragged the United States deeper and deeper into the war.

Bill Reeder felt the patriotic burn so many young men felt for their country at that time. He was young and had no other serious direction at that time. And, lest we forget, it was a popular war early on.

He enlisted in the U.S. Army in August 1965 and went to boot camp at Fort Polk, Louisiana.

He did well, and he liked the Army. From Fort Polk, he was sent to artillery school at Fort Sill, Oklahoma. When afforded the opportunity, he went to Officer Candidate School at Ft Sill. He was commissioned as an officer, a Lieutenant of Artillery, in August of 1966.

Then Lieutenant Reeder was sent to Fort Carson, Colorado, with an Honest John rocket battalion *for a year.

Again, he did well, but the adventure of aviation was calling and it was too much to resist. He got into flight school and was given a "Mohawk" transition (trained in the Mohawk aircraft).

---

* The **MGR-1 Honest John** rocket was the first nuclear-capable surface-to-surface rocket in the United States arsenal according to https://en.wikipedia.org/wiki/MGR-1_Honest_John

OV-1 Mohawk

The Mohawk, nicknamed "The Widow Maker," was a fully aerobatic, twin-engine aircraft used primarily for intelligence gathering missions. Mohawks were actively flown by the military from the 1960s to the early 1990s*.

Lieutenant Reeder commenced his first tour in Vietnam, serving as the Operations Officer in a Mohawk unit. Mohawk missions in those days were exciting and risky.

While there, he flew something in excess of 200 missions.

Then he was shot down.

---

* **Grumman OV-1 Mohawk**

This is a light attack and observation aircraft developed by Grumman Aircraft Corporation. It was designed for battlefield surveillance and light strike capabilities, featuring a twin turboprop configuration and carrying two crew members in side-by-side seating . The aircraft was intended to operate from short, unimproved runways in support of the United States Army maneuver forces. The prototype (YAO-1AF) first flew on April 14, 1959, and the OV-1 entered production in October 1959 . The aircraft had various variants, including the OV-1A, OV-1B, OV-1C, OV-1D, and others, each with different configurations and capabilities. (Source: Wikipedia)

REEDER DIDN'T GET to enjoy the luxury ejection device that was later designed and used in Mohawks. The later model Mohawk had a much safer ejection seat that was a self-righting, rocket propelled device that could be initiated at zero-zero, an aviation term for the moment right before the aircraft begins to fall*.

Would that he should be so lucky! No.

When Reeder ejected, he was using an early version. He was propelled through the canopy by an explosive device that looked like a 40mm round. It was sort of like a giant .22 caliber short round except it was about two inches thick and about six inches long.

To operate it, you just pull the handle and BANG! You were suddenly flying through the air, hoping you would remain attached to your seat and chute, praying you didn't receive a back injury from the explosion or a flailing injury from the wind blast.

It was not as bad as it sounds. No other Army aircraft had any ejection device at all.

———

FOR STARTERS, he did survive the ejection.

Now, he just had to survive hiding out until he could be rescued.

Speaking of that event, he modestly recalled, "I got shot down in an area where, until me, they had *never* rescued anyone. The unit I went to lost 16 of its original 18 aircraft. I was only the fourth crewmember to be recovered at all, among 18 aircraft and 36 crewmen."

---

* Zero-zero refers to zero up velocity and zero forward acceleration. Zero up velocity means that the aircraft had not begun to descend/fall yet. This is the moment right before the aircraft begins falling from the sky.

Reeder casually added, "Anyway, after my bail-out and rescue, I spent a short time in the hospital undergoing physical therapy. Then it was back to duty."

———

HE MADE IT BACK HOME, and he started going to college at the University of Nebraska in Omaha. He earned a degree in Political Science on the Army's dime.

However, he had known since his return home that his next military move would be a direct assignment back to Vietnam. He was looking over his shoulder, so to speak, for that to happen.

He was invited to be to be a Mohawk instructor pilot, which would been an envied position, out of enemy danger.

What Reeder wanted, however, was much more dangerous. He asked instead for the helicopter qualification course. The Army was only too happy to oblige. There were pilots all over the Army waiting in line for Mohawks, but there was a great need for rotary wing (e.g. helicopter) aviators.

It was, Bill said, a move he never regretted. He completed the rotary wing qualification course, and then Cobra (AH-1 attack helicopter) training*. The normal mission for the Cobra was to place fire on the enemy while they were engaged (fighting) with friendly ground forces.

———

* The Cobra model Reeder flew in 1972 was commonly mounted with a 7.62mm mini-gun, a 40mm grenade launcher and 2.75-inch free-folding aerial rockets. It was a tandem seat (pilot sitting behind the copilot), narrow profile (three-foot wide) gunship, originally rushed into production to fill the air-ground support role. It was created with standard Huey parts. For example, the transmission, tail rotor system and engine were all Huey parts.

2

BACK TO VIETNAM

CAPTAIN REEDER BEGAN his second Vietnam tour in December 1971.

Captain Reeder standing in front of a Cobra. 50 caliber bullet holes are visible in the helicopter canopy (windows) behind the two men.

That year, the Vietnam War had become a nightmare. The South Vietnamese were suffering. American leaders were trying desperately to find a way out of it, and there appeared to be no end in sight.

The political landscape was fraught with a failed war in Korea. The Army couldn't seem to find the political will to win in Vietnam. This was made worse by the activism breaking out at our nation's colleges and universities.

There were public and private negotiations under way to end the war and bring our boys home. The talks had gone nowhere, but the nation hoped the end was in sight.

Half of Reeder's Cobra class had their Vietnam orders cancelled and were sent somewhere else.

"We all thought the war was over," Reeder recalled. "We thought we would mop up and maybe not even finish our full 13 months overseas*."

WHEN HE ARRIVED IN VIETNAM, the local commanders wanted him to take a leadership promotion. In fact, this happened repeatedly. At every headquarters unit he passed through on his way to his assignment, he was asked to accept a leadership job. Captain Reeder politely declined and pointed out that he had *just* become helicopter qualified.

He asked to be sent to a line unit where he could use his skills and gain experience in the cockpit. He was told there were two units available just then. One was the 2/17 Cavalry at Camp Hollaway. The other was, an "attack unit with a bunch of rowdies and a strange special operations support assignment."

This latter organization flew mostly classified missions in support of military special operations, and Reeder jumped at

---

* "In fact, when I arrived in Vietnam, since forces were being pulled out [and there were fewer experienced officers being transferred in] it was very hard to get assigned in a line unit down to tactical level."

that assignment. He was already experienced in special operations missions from his Mohawk days.

He joined the 361st Aerial Weapons Company, A.K.A. the *Pink Panthers,* exclusively made up of Cobras. He took command of the 3rd Platoon. His fate was sealed.

Things were a bit slow at first. He got to support long range patrol operations, which saw their fair share of action.

In April, the Communists launched a three-pronged offensive and swung across the demilitarized zone with conventional armored forces out of Cambodia. They also did a sweep into Saigon and into the Central Highlands.

It began slowly in mid-April with the over-running of some outlying fire bases.

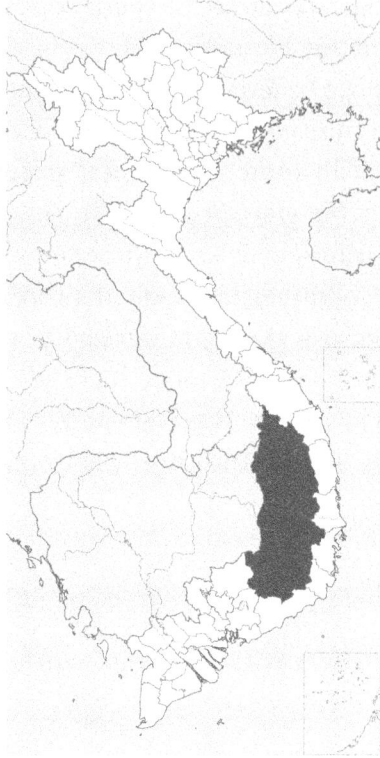

Central Highlands of Vietnam, as found on Wikipedia Commons, image By TUBS

Then suddenly, on April 24th, in a period of about two hours, the Communists overran all the friendly forces at Dak To and Tan Kan. This offensive was a massive, unexpected victory for the North Vietnamese Army.

8th Infantry Rgt. descending Hill 742 during Operation MacArthur 1967, By US Army. 5 miles from Ben Het

After Dak To and Tan Kan fell, everything in the Highlands fell (except a few towns and a Vietnamese Ranger unit at Ben Het in the tri-border area of Laos, Cambodia and Vietnam.)

The only friendly places left in the entire Central Highlands were Pleiku, Kontum and Ben Het.

With the April 24th blitz, The 361st cancelled its support of Long-Range Recon Patrols. They shifted to fully supporting the South Vietnamese and American Advisor forces in the Central Highlands. The only American units remaining were American aviation units and American advisors to Vietnamese units.

The War hadn't wrapped up. It had reignited.

3

SHOT DOWN, AGAIN

ON THE MORNING of 9 May 1972, Captain Reeder was scrambled out on a Tactical Emergency (TAC E) to support the fight at Polei Klang.

NARA photo by SP5 Gilley, via Wikimedia Commons, Ben Het Vietnam

His copilot (front-seater) was a new arrival in Vietnam, a First Lieutenant named Tim Conry from Phoenix, Arizona. Conry had been in-country about two weeks. Reeder had taken him under his wing since his arrival, and they had flown together almost every day.

The report they received was that there were "tanks in the wire," meaning enemy armor was breaching the perimeter. Their mission was to escort a highly classified unit, call sign "Hawk's Claw," UH-1 ("B" model Huey). Hawk's Claw would knock out tanks with wire-guided missiles[*]. The mission went well. The Huey expended one load of missiles, and they returned to re-arm and refuel.

During the re-arm operation, they got a call that the ground guys needed ammo badly, especially anti-tank ammo ("ordinance"). There was a slick (a UH-1 Huey helicopter, with no armament, used to haul troops and cargo) coming up with ammo to re-supply the ground forces at Ben Het.

"The American advisor had asked us, and we agreed to support the ordinance ship with cover into Ben Het. I had been lead on the morning mission, so we sent the Huey[†] to hold the east. We went in to support the supply ship."

They met the supply ship and escorted it into Ben Het. When they arrived, the situation was serious. The enemy not only had attacked with tanks, they had breached the perimeter and occupied half to two-thirds of the firebase.

Now this was a typical Ranger Special Forces firebase, about a half mile across. There was lots of activity in the camp and there was lots of anti-aircraft fire, some 23mm and some .51 caliber.

Reeder explained, "When we got close, we elected to come

---

[*] That system was the prototype of the modern TOW anti-tank wire-guided missile. It proved to be very effective and very accurate.
[†] Hawks Claw, the classified anti-tank bird

in low and fast, rather than descend from a thousand feet with bright clouds silhouetting us. We flew staggered to lay down covering fire on either side and forward of the resupply ship."

There had been no thorough planning meeting for this mission. It was the chaos of war. The defenders called for ammo; the supply guys called for Cobra support. The Cobras met the slick and had successfully cleared it into the pad at about 0930 hours. Then the UH-1 pilot changed the program.

After he had kicked out the ammo, the pilot elected *not* to take the pre-planned route out. Instead of flying out in the same direction he was already going, he did a 180° pedal turn, that is, he turned around on the pad and flew back out the way he came in.

Tactically, this change meant that the Cobras could not simply fly through the fire once, past the base and out of danger. They now had to turn above the firebase to try and support the Huey with suppressive fire. As soon as Reeder began the turn, he started taking fire from two .51 caliber positions.

Hits shook the aircraft, moving from aft to front. It became immediately apparent that they were going down.

He made a radio call to his wing while his gunner tried to suppress the fire. But the gun emplacement was out of reach-out the right door, aft of center, and just outside the traverse of the turret.

"The first thing we lost was the tail rotor. Next, the engine. The impacts sounded like someone was standing on a work platform outside the aircraft hitting it with a jack hammer. Then it got all quiet," Reeder recalled.

They were on fire, descending in a spin.

At the bottom of the autorotation*, they experienced a

---

* Autorotation is a condition of helicopter flight during which the main rotor of a helicopter is driven only by aerodynamic forces with no power from the

partial hydraulics failure. There were no "hydraulics out" lights, but it proved very hard to get the collective up to cushion their landing.

"I don't remember too much on the bottom side of that crash landing," Reeder said, "but later, I heard from several witnesses."

He eventually got the story from the airmen in the aircraft at his wing and one of two guys watching from the ground. They said the aircraft descended turning and burning. Engulfed in flames, it hit nose low on the left side. It bounced up and did another couple of turns. Then, it settled upright on the skids.

This was a miracle for the Cobra, which usually landed on one side or the other. If it had landed on its side, the only way to escape for both men would have been to use special tools to break out the window canopy, called break-out knives*. If the Cobra had rolled onto one side or the other, Reeder and his copilot would not have lived.

---

engine. It is a manoeuvre where the engine is disengaged from the main rotor system and the rotor blades are driven solely by the upward flow of air through the rotor. In other words, the engine is no longer supplying power to the main rotor. A vector of the rotor thrust in a helicopter is used to give forward thrust in powered flight; thus, where there is no other source of thrust in a helicopter, it must descend when in autorotation. Autorotation is a means by which a helicopter can be landed safely in the event of an engine failure. Source, https://skybrary.aero/articles/autorotation SKYbrary, *autorotation* entry

* A break-out knife was a knife with a thick, sharp blade less than two inches in length with a heavy metal handle that enabled a pilot to strike a Plexiglas canopy and break it out. The Cobra is a tandem seat attack helicopter with a narrow front profile. The body of the aircraft was only about three feet wide; it was top heavy when sitting on its skids. It was designed to provide a minimal sized target opportunity to the enemy when it rolled in directly on a target. The cockpit was odd in that the front seater door opened left, and the back seater door opened right. In this incident, both crewmembers in those seats were injured, so if the aircraft had landed on its side, we speculate that neither would have been able to break through the plexiglas canopy with only break-out knives to escape.

"I do remember the impact," Reeder went on. "Smoke and flames immediately came up around the cockpit."

The last thing he remembers is that after he yelled to his front seater to get out of the aircraft, his front seater responded in the affirmative and departed.

4

CAPTURED

REEDER WAKING to find his body on the ground was proof enough that he did escape the aircraft, but how he did not know. He would strangely find out more about how it happened years afterward.

In 1981, he met the American advisor who had *seen* his craft crash land from the Ranger camp perimeter. He was a soldier named Steven Mark Truhan from Rutland, Vermont. The camp was up on a hill, and Truhan could see the aircraft crash just outside the wire.

He saw the front seater open the hatch and exit the area, and then he saw that the back seater open the right hatch. He saw someone start to get out of the helicopter. This second man caught his foot on something and fell, leaving him hanging upside down outside the right side of the aircraft.

"[Truhan told me] the aircraft was burning. He didn't want to see a guy hang there and burn to death."

Truhan raised his rifle and put the man in his sights. He was just ready to squeeze off a round and prevent that torturous death when he lost all visibility. Smoke completely covered the aircraft. He couldn't see anything.

Reeder said. "The smoke was there just a few seconds, and when it blew clear, the guy was gone."

Truhan looked for that pilot for years, and he finally located Reeder in 1981.

Reeder continued, "That really got my attention. Throughout this experience, in two tours in Vietnam, I came very close to dying many times."

THE NEXT THING Reeder actually remembers was waking up in the weeds. He had gotten off the helicopter and moved some distance away and passed out there. He is not sure how much later it was, but the aircraft was still cooking off rounds.

He was extremely disoriented. He knew he was seriously injured and in a lot of pain. At first, he was totally paralyzed. He was too groggy to comprehend how bad his condition was, but he was really baffled about how he could have moved out of the helicopter and later be paralyzed.

In reality, he *had* broken his back, but he had not severed the spinal cord. The explanation his doctors later offered was that when he crashed, he could still move. Then, the break acted like a bruise and swelled up, causing the temporary paralysis.

He lay there for several hours, awake and unable to move. Gradually sensation started coming back; his arms first, then later in the day, his legs, so he was finally able to crawl around.

"The first thing I did was get my radio out," Reeder said. "I was in such a state I was not sure where I was or what had happened. In fact, I was mentally floating between this current incident and my previous Mohawk shoot-down in my first tour. At times I thought I was on the ground, that I had just been shot down out of a Mohawk. But as time went on, that day, things gradually became more clear in my mind."

The radio didn't work. It was an interim radio; the kind that could come on in the vest. It had a faulty switch on it. Also, the battery was dead, and there was no spare. He used his signal mirror the rest of the day, to no avail.

Reeder did eventually get up and crawl around looking for the front seater, but he couldn't find him anywhere. He suspected the lieutenant's radio was working, and that he possibly had gotten away already. So, Reeder's concern for the other man wasn't much.

"Around dusk, a flight of helicopters came in; a bunch of Cobras, and I couldn't see them but I could hear them as well as some other helicopter*. They swooped low, dropped down [landed], and started to come out. I made the assumption at that time—and I held it the whole time I was a POW—that they had been in radio contact with [my front seater] and he had been picked up.

"Actually, he was either dead when they picked him up or died on the way to the hospital," Reeder continued. "I never did establish whether they had radio contact with him, but they had seen him all day. He had gotten out of the helicopter and moved over to a little bombed out area and lay in a crater. They had seen him in that crater."

Reeder hadn't seen him and did not know the copilot was there. Reeder had apparently moved away from the copilot while looking, and was nearly immobile.

"When I saw the aircraft coming in that night, I got my

---

* Every model of helicopter has a distinct sound. As in, so distinct that anyone even vaguely familiar with that craft can identify its sound from thousands of feet below. Therefore, Cobras sound distinct from Hueys. Apaches sound distinct from Black Hawks. Reeder here identified Cobras by sound and 'some other helicopter" meaning one that he didn't immediately recognize from the sound.

strobe light out," Reeder said. "We had a little blue cover that we put over the strobe light so it wouldn't be mistaken for ground fire. In fact, [the flight departed] almost exactly toward me. Well, the friendlies saw the light, but even with a blue filter, they thought it was enemy fire. The front seat in the lead Cobra put out a stream of mini-gun right on me. It seemed to me it was two feet from me, and I rolled out of the way."

Reeder later actually found and spoke to the man responsible for shooting at him. Gordon Bently Hill remembered the whole incident. He thought it was ground fire.

"There were no hard feelings anyway," Reeder said with a chuckle.

He went on, "I was just outside the wire of the camp. Between me and those friendlies was the clear no-man's land [an open area cleared with agent orange and bulldozers]. In crossing that open area, I would have had to go through attacking Communist forces, get across the open area across concertina wire, through claymore traps and a minefield, pound on their command bunker and say, 'Hey! I'm here! Let me in!'

"I didn't think that was a viable option at all, and it was frustrating to be a few hundred yards from friendlies and to have no access to them. Talk about being so near and yet so far. But there was no way I could get through that mess.

"There were only two other options. I could go back to Kontum, but they were completely besieged by enemy. Or I figured I would have to go all the way back to Pleiku, which I believe was about 30 miles. At that point I could only crawl, but I thought I could certainly make it 30 miles.

"I knew I could not stay where I was. That was the other consideration because as the sun set, all kinds of air strikes were coming in. There was an AC 130 working the area with his 20mm, and I was in the area where all the bomb strikes would be laid to drive back the enemy. Though they were concerned with pilots, they were not going to save the life of one guy to cease bombing an area and risk losing the camp.

"So, I moved out to the southeast away from the camp.

"I traveled all that first night, crawling mostly. I was crawling for a while, then hobbling with great difficulty. As I moved away, first there were woods, then heavy jungle. On the first night, bad guys were everywhere, screaming and shooting. I was in the middle of a major assault. I guess I was lucky I was moving at night.

"By the morning, I had moved into the rear of the assaulting forces and felt a little more secure. Maybe I was lucky I had to crawl that first night, but I had so much difficulty. I decided to try moving in the daylight after that.

"I was thinking back on all the survival things I had been taught, which at that time in the Army was not much. We had a few days survival training during flight school, and I had gone to a survival school in the Philippines before my first tour.

"There I learned that I should stay off trails, that I should travel at night. Those types of things. But it was too difficult. I ended up moving in the daylight.

"Then I had a series of disappointments trying to get picked up. Nobody came in the first day because on the first day, a conscious decision had been made *not* to pick up anyone. This was very unusual, but it was because the enemy in the area was so vast and the anti-aircraft fire so extensive that any rescue would be suicide.

"The guys who picked up my front seater, true to the craziness of Army aviators of the time, got authorization to do an armed recon in the area of the assault. They had planned all along to make the rescue against all orders.

"Next day I heard a Vietnamese Bird Dog [a small, fixed-wing scout airplane] overhead. I came to the edge of a field with jungle all around it. I guess it was 40 meters across, and I fired a pen flare.

"He did not see it, but the bad guys saw it. I could see them across the clearing, firing their weapons and coming quickly. I was in my hobble mode at that point. I hobbled back toward the jungle down a little trail, moving along the trail just to get out of the area quickly. I got across a stream, and coming up the other side I could hear these guys closing behind me. I got around the bend and rolled off the trail, almost like in a cowboy movie.

'They came running along the trail shooting and screaming. They ran past.

"Survival school was not very helpful for me at that moment. I knew I had to stay off the trails. I knew I should try to move at night, and those types of things. I tried to apply those, but I gave up moving at night. It was just too tough.

"I rested that night, and there were B-52 strikes all around me, just rattling the earth. On the third day, I got up and continued my movement to the southeast off trails just through the hillsides.

"While moving through a cratered area, I saw an FAC [Forward Air Controller, the Air Force spotter plane] overhead, moved into the middle of the bombed-out area, and took off my Nomex flight suit shirt, waving it around in hopes that he would see me.

"Indeed he did."

Reeder saw the aircraft make a couple of circles and fly away. The next thing he expected was a helicopter to come in and pull him out, but quite the opposite happened.

"After several minutes, what I saw was the FAC returning with a flight of F-4s and putting them on my position. So, I jumped in a crater and hugged the side. Luckily F-4 bombs don't fall that accurately.

"Later that day, as I was walking down the side of a hill, I quite suddenly heard a bunch of alien voices in front of me. I couldn't speak any Vietnamese at that time. I hoped they were

Montagnards [another peoples friendly to the US Army], but I didn't know.

"I froze in my tracks and slumped down hoping to be mistaken for an animal or whatever. But that was not to be the case. These guys were very close to me and within just a few seconds I had five young NVA (North Vietnamese Army) regulars armed with AK-47s pointed at my head."

5

HUMANE AND LENIENT TREATMENT

"I WAS NOT ARMED," Reeder confessed.

Even though he had been issued 38s, pilots felt they were poor survival weapons. Reeder had been content to carry CAR 15s in his Cobras behind the seat. He also had had two bandoleers of ammunition.

"There is an old adage I never listened to- but I certainly do now.

"'If it ain't strapped to yer body, you ain't getting it out of the aircraft.' So, I did *not* get my CAR 15, I didn't get my bandoleers and ammo, I didn't get my survival kit. All I had was in my flight vest."

"I honestly thought I could evade capture, I really did, the whole time.

"I was so totally disappointed. Actually, disappointed does not begin to cover the feeling. Your entire support structure evaporates the instant you are standing there with a gun pointed at you, and you are at their mercy."

"The jungle was so thick that I had come to within 10 feet of these guys without even knowing it. They had me stand up, and they brought me forward. Just a few feet further the area was cleared. I saw a stream where these guys had been collecting water."

He was led up that hill to a large bunker complex. All the while he thought he was escaping, he had been staggering toward a bunker complex, the staging area from which all the enemy assaults on Ben Het had been launched. They kept him there, during which time he endured the most brutal three days of his capture.

"The guys who captured me weren't too bad at all. They simply tied me up and left me for the interrogator, who came much later in the afternoon."

THERE WERE a lot of guys being lost and captured in those days, and many soldiers would talk about saving the last bullet for yourself. Reeder said he never quite subscribed to that philosophy. He always believed in the value of life under any circumstances, even torture, was worth saving. He held that living meant the hope of surviving to return to a normal life. This was probably worth taking anything the enemy had, rather than putting a bullet in one's own head.

REEDER MADE it clear that the training he got in resistance did not help during the three days he was held here. The training stuff he got was based on E&E (escape and evasion). If the trainee did not get captured during the training, he did not have to go into the POW training camp.

Reeder was never 'practice captured' in training, and he did not have any of the POW camp training. The Code of Conduct, on the other hand, was very valuable. He noted that he would have preferred more information on the Geneva Convention, but the Code of Conduct was crucial.

The interrogator arrived and started with general conversation, and since he spoke English, Reeder told him that his back hurt terribly.

The interrogator had probably figured that much out anyway because the lieutenant had been wounded in the ankle and had blood all over his face and head. For three days he had no bladder control; something he believed had to do with the temporary paralysis. He knew he looked- and smelled- terrible.

"He [the interrogator] registered all this information with a gleam in his eye. Then he got into tactical questions I would not answer. I gave him only my name, rank, serial number and date of birth and minimal discussion on my health."

The interrogator decided to get as much out of Reeder's pained back as he could. The man had Reeder's arms trussed behind him until the elbows touched, which was very painful. The captive was then tied to a tree in a very straight position after all this time hunched over with a very painful back.

The interrogator proceeded to beat on his captive while asking questions Reeder would not answer. That continued all afternoon, until it was dark.

At night, Reeder was thrown down into an abandoned bunker with six inches of water, with bugs and everything else. He was kept in there until morning.

"This continued for three days. I felt quite proud that I did not give any information. This was not heroic, because the more brutal they are, the more you are at loggerheads, but I had no trouble resisting."

On the third day, Reeder's interrogator said words to the effect of, 'We have tried to work with you the past three days. We have asked only that you show us some cooperation so that we might show you the deserved humane and lenient treatment of the Vietnamese People. You have chosen not to cooperate with us in even the slightest form. We have many operations going on and have no more time to devote to you. If you do not cooperate, you will be executed.'

As the Vietnamese man looked back over his shoulder, Reeder saw a dude with an AK-47 rifle.

"At that point, an old training film came to mind, where this scenario was played out in the Korean War. In that film, a Korean interrogator threatened to shoot a prisoner with a pistol if he did not cooperate. Then he pulled the trigger and nothing happened. Not that I did not believe the gun was loaded, and that I would survive, but there was no way I was going to give them any information. Whether I cared if I died or not did not enter into it. I guess I was just numb, and I was probably not as frightened as I might have been.

"The guy pointed the gun at me. The interrogator said, "This is your last chance." The guy squeezed the trigger. And nothing happened.

"After that the interrogator left in disgust and told me to go. One of these guys gave me a card printed in English. It was a little larger than a business card.

"It read, 'You are being taken by representatives of the Peoples Liberation Army. You should be cooperative. The intent of their taking you is to take you outside the range of American artillery and American bombing strikes where you can receive the humane and lenient treatment of the People of Vietnam.'"

6

THE BAMBOO CAGE

CAPTAIN REEDER WAS GIVEN BACK his boots without socks or laces. They gave him a little bag with a couple of rope straps; in it were several pounds of uncooked rice. It turned out to be his own rice rations for this trip. All this occurred about six days after his shoot down.

He did not know the extent of all his injuries. He had a broken back from the impact of the crash, an ankle wound, some superficial head injuries, and some minor burns on his neck. He still had no bladder control, and it was very irritating to him and his captives. He was unshaven and had various leeches on his body. In fact, one of guards laughed because he had a leech up his nose.

"We set off. The only communication, and almost the only words in English they knew were, 'Go quick or die.' The guards were armed with AK-47s with bayonets. We traveled one day to the south and two days to the west."

They passed through what was obviously a major battle zone. They did not pass too many reinforcements moving forward, but they did pass many military evacuees moving back (wounded or regular North Vietnamese Army soldiers).

The wounded NVA were transported any way they could be; often their litter was actually a shelter half on poles. Most of them were in extremely bad shape, and when the POWs came into contact with them, they would scowl and spit and yell.

Reeder's condition continued to deteriorate. He began getting blisters the first day from walking in boots with no laces and no socks. By the time the three-day trip was finished, his feet had more blisters than ever before.

He was very tired, and there was not much water. The first night was spent on a sort of campout in the woods. There were a lot of other Vietnamese forces in the area. They built fires and covered them the best they could with vegetation. They boiled rice and ate that, then moved on.

The second night was spent in a regular way station along a major trail. The Americans suspected that by that time, they had actually crossed into Cambodia.

During the first and second day, they suffered some pretty extensive air strikes. All along the trail network the enemy had little bunkers along the sides of the trails. When the aircraft rolled in, they would dive into these little bunkers. The whole world would blow up. The aircraft would leave. They would crawl out of the bunker and continue on the journey.

The third day they arrived at Reeder's first prisoner of war (POW) camp. When they arrived, they moved through very thick jungle before coming out into a cleared area. It was actually a bamboo thicket which was used for construction material for the camp. It looked like an old stockade from American history, but instead of logs it was made of bamboo.

It was constructed with two stockade perimeters, one about six or eight feet inside the other. Dug out between the two perimeters was a sort of moat all the way around. This 'moat' was a dry, empty [ditch] with sharpened (bungi) sticks pointing up from the bottom.

The enemy used a flattened log bridge to get across the 'moat' and into the camp.

"Now here I come, after all I have been through—my capture, my interrogation, the brutality, the movement over three days along this trail. I did not know if I would make it. Somehow, I got across the log and into the camp," Reeder recalled.

It was a fairly large camp. He estimated that it was 150 yards long with a bunch of bamboo cages. The first thing they did was relieve him of his little rucksack and the remaining rice.

"I was treated not exactly violently. I was just treated rudely and put in a cage."

His cage was 13 feet wide, 20 feet long, and four and a half feet high [not tall enough to stand up in]. The cage had 25 or 26 other prisoners in it, packed in like sardines. Down the middle of the cage was a log, split with holes cut in it, like a huge stock for the prisoners to put their feet in.

A.I. rendering of Bamboo Cage as described.

His boots were taken away from him, and his feet were "just an awful disaster!" They were bleeding badly. Also, Reeder was just getting control of his bladder.

Reeder says, "I am sitting there, wondering what the hell was going on."

He didn't have to wait long.

In just a few minutes a Communist guy approached Reeder with with a chisel and a hammer. He grabbed his foot and looks at it.

Cold fear gripped him as he realized the man was intending to gouge the bottom or his feet to disable his ability to run or escape.

The man inspected his foot, put it down, looked at the log, and called in a guard with a rifle to watch the prisoners. He raised the log and started chiseling out a hole.

Reeder could now see that the holes were made for Vietnamese feet, but they were too small for an American. The man was cutting the hole bigger to fit his feet.

Reeder tried to talk to the prisoners around him, but they wouldn't answer. At first, he thought they did not speak English. But later in the evening, when there was minimal guarding, he learned that most of them did speak English. They hadn't answered him earlier because they were under strict orders from the Communists not to talk with *him*.

"So that was the situation in those cages," the American said. "I lived like that for a couple months. The conditions in that camp were so bad that I don't know how long I could have survived there.

"We got to eat two grapefruit sized balls of rice a day. Rice that had been cooked and pressed into a ball. Then these balls were passed around to all the cages and given to all the prisoners, once in the morning and once in the evening."

Captain Reeder was interrogated every day in the jungle

camp. Even in those Cambodian cages they gave him paper and pen and told him to write what he thought about the war.

"I was interrogated every day in that camp in the jungle. Every one would start the same. They would pull you out of your cage. They set you down with your interrogator. They give you a cigarette and let you light it up. They would offer you some tea and start asking questions.

"And they would start being belligerent if you did not cooperate. But I never had any problem with that. I didn't tell them anything."

Reeder said he would think better of it now, but he wrote pretty well what he thought then. When they read it, they would be angry.

"You not tell truth! You not write truth."

Then they would make promises. Something to the tune of, "Hey, if you will write the truth and just cooperate with us, you will get better food, medical care, and we'll let you write home. You will be able to receive letters.'"

His captors got very little out of Reeder, but he had more urgent survival issues to worry about.

"I don't recall water being a problem in that camp. They would have pieces of bamboo filled with water in the morning, and this piece of bamboo would be passed to all the prisoners."

But the latrine facilities were awful. As Reeder spoke about it, even his face showed it. This must have been where most of the disease in the camp came from.

Every day at about the same time, the guards would let the prisoners out of the cage for about five minutes. Having to plan his need to fit the guards' schedule, each prisoner would go over to the latrine facility, which was nothing more than two holes dug over at the edge of one tent with a small cot. Then he would dutifully squat over a hole, take careful aim, and let fly.

There were two major problems.

First, to get to it, the prisoners had to walk through 10-15 yards of human waste piled a foot high. Then they could do the deed, but they had to walk back through it after. Reeder's bare feet- badly blistered- and his ankle wound were constantly exposed to the stuff. There was no way to wipe or wash it off. It was tracked back to the cage where dirty feet and legs were locked back in the stocks.

The second problem was the men who were already very sick. "There were a lot of guys in the camp with dysentery," Reeder said. "The sickest ones were almost dead. They were put out in hammocks over next to the latrine. There was no worry about their escaping. They could barely move. When they tried to get to the little hole, they rolled out of their hammocks, take a few steps, and then crap all over the place with this terrible diarrhea."

Of approximately 300 prisoners in that camp, all were South Vietnamese except for Reeder and one other American. NVA were carrying out bodies every day.

The recently renewed offensive had overrun most of the fire bases in the Highlands. That's where all these prisoners had come from, Reeder recalled.

"By the time I was captured, hell, the only things left standing in the Highlands were Kortum City, Pleiku City, and the fire base at Ben Het. All of these [held their ground,] at least in the short term, so there was no more massive acquisition of prisoners after that point." Reeder had been one of the last to come into the camp.

After a few weeks Reeder was moved to a smaller cage with fewer prisoners, where he met the other American. He was Wayne Finch, OH-58 scout pilot of 2/17 Cavalry. He had been shot down a couple of months before Reeder and was in better shape. His injuries were mostly burns, which had pretty well healed, and he had adapted to the diet of rice.

The prisoners learned that it is hard to eat a large quantity of rice, but Wayne was eating all his rice. In fact, he was well enough that they sent him out on work details, giving him a bit of exercise. Reeder was not so lucky. Between his back injuries and his infected ankle, his condition was deteriorating daily.

"Wayne and I discussed escape and shared some ideas on how to get out of the camp. We knew we couldn't go over the top of the fence but there was one little area where a stream went through where they got water, and we were thinking we might be able to get under the fence.

Eventually, Wayne and Reeder concurred that Reeder was in such bad shape he couldn't survive more than a day or two, even if they got outside the camp.

"You get some strength back, and we will try this thing later," Wayne said. So, they did.

Only one major escape attempt occurred while Reeder was in the camp. A group of Montagnards, all in the same cage, broke out and tried to climb over the stockade wall.

"A lot of them were shot trying to get over the wall," said Reeder. Those who survived were tied with their elbows behind their back. They were then strung up in various contorted positions from a few trees in the camp.

The Montagnards were left up there for at least a couple of days, just hanging there, to the guards enjoyment. Then the NVA cut down the few still alive and put them back in the cage.

Then the problems of dysentery kicked in. Both Americans experienced attacks of dysentery. The situation was complicated with the unbending routine of the latrine schedule. There were no exceptions to the daily five-minute run to the latrine. Other than that, they did have a piece of bamboo to use as a urinal. But with dysentery, when the prisoner had to go, he just went all over himself.

"It was really a god-awful situation," Reeder said, grimacing at the memory. "It is hard to describe how it worked. They were carrying these Vietnamese bodies out and burying them. A lot of them were dying from dysentery, and a lot were dying from malaria.

"Wayne made a comment in those days when we were in that camp. 'Geez. These people live here; this is their country. If they are dying from it, If I ever get malaria, that's it. I'm done for.'

7

MARCH TO HANOI

ON 2 JULY 1972, the Communists came and pulled certain people out of cages. Finch, Reeder and one other man from their cage were brought outside.

There was a group of people standing there. There were 25 South Vietnamese officers and the two Americans. All had their arms tied loosely behind their backs, then tied together.

Reeder described the scene.

"They told us they were going to move us to a new camp. They said it would be a better camp. Conditions would be better. There would be better food. We would get some medical care. We would be able to write and receive letters. They said we should try to make it. No mention was made of what the hell happens to those who did *not* make it. They said it would be a long way, and it would take up to 11 days to get there. They did not give us our boots back this time. We went off barefoot.

"I came to know most of the Vietnamese in that group," Reeder said. "We thought we were going to another camp in the jungle somewhere; a better location."

In fact, the trip lasted more than three months and ended up in Hanoi. Out of that group, seven of the South Vietnamese died. Wayne died. There were eight deaths in total, mostly from dysentery or malaria. Wayne ended up dying of malaria. He made it marching for over three months, just to die five days out.

Reeder realized by now that he had to have a goal; something to give him the will to get up and go each day. His condition worsened by the day, and he found the general desire to go home was not enough. Everyone wanted to go home. Freedom and return to the United States was the ultimate dream of all.

In his words, "But when you look at how to survive to see the sun come up the next day, the real burning goal I had was to get to Hanoi," Reeder confided. "I knew after a while that we were obviously not going to some camp in the jungle. It was obvious that we were going north. Even some of the guards by that time let down the charade.

"I figured that if I could get to Hanoi, I could probably get into some kind of stable situation. Perhaps a regular prison where I could get, you know, maybe a little gruel or some kind of regular food. Maybe I could get some medical care and be done with the agonies."

He expanded a bit on the agonies— the sum total of the anguish of living under tyrannical, torturous, brutal men after being seriously injured in an aircraft crash-landing.

"You cannot imagine what it is like being in so much pain, suffering so badly from disease. I started with the injuries I had when I was captured, then add dysentery. Then, I also got malaria. I got three types of intestinal parasites.

'My leg wound got so infected it was swollen two to three times its normal size. It had cracks in the skin all over the leg. The color of the leg was a blackish-bluish green color, and there was pus oozing out all over the leg. So, with that kind of pain, and being that- *uncomfortable* is not the right word- being that miserable, it cannot be described."

As Reeder struggled to describe this situation, he repeatedly searched for the word that would illuminate the reality for a listener or reader. Or maybe a soldier trying to study the lessons here.

"You come in at the end of the day's march in the jungle. You lie down and try to get some kind of sleep. And at least they gave us a sort of hammock to sleep on, off the ground."

Then almost to himself, he began to mutter as he recalled the moments maybe even *he* can't believe sometimes.

"But I guess we wouldn't have survived lying on the jungle floor with the bugs and snakes and all kinds of stuff if they hadn't."

Then he perked up a moment and went on.

"How do I describe the emotional feeling?

"There, to be awakened in the morning, to know I had to get up off that hammock. Why, just rolling over was very painful. But standing up. The most excruciating pain I can recall in my life was standing up in the morning: having the blood rush back into my leg and just this... this..."

He stared at his leg, shaking his head slowly.

"Well, you know how the throb feels when you pinch your finger badly, like in a door. Just the throb of that and your nerves go frazzled. You can think of nothing else but pain."

He smiled that smile of his, when he knew he could not describe it. Regardless, he went on, determined to stab at it some more.

"I can never remember feeling more pain than that. But you force yourself to pick up the little rucksack and stuff, put it on your back. And, then you head on out."

In addition to injuries and wounds, all that walking barefoot through the jungle on trails of rocks left the bottoms of his feet in shreds, with cuts all over.

"I can never fully relate what it felt like to go through that every day. You are continuously just crying inside, just whimpering inside. And hoping you will make it to the end of the day.

"We were walking one day, along a damn creek bed. Then we had to climb along a cliff next to a waterfall, then get up and continue to march. I was hobbling through this creek bed, barefoot. It was not nice smooth pebbles like you might expect. It was a bunch of sharp rocks under this water.

"I said to myself, 'Damn! I don't ever want to hear a woman complain about having a baby again. No matter how bad it is, it can't be this bad.'"

He laughed after he said this.

He brightened then, and he was seemingly past this particular painful memory as he smiled broadly.

"I can remember that I was trying to maintain my sense of humor. I believe very strongly that in survivors, a sense of humor is required. Once I would get going, I would put the strangest things in my mind and just keep going.

"I would imagine I was hiking along this make-believe trail. I called it the California Riding and Hiking Trail. I would be strolling along the California Riding and Hiking Trail. I could imagine these nice painted white rails alongside it, beautiful hard packed trail, some leaves freshly fallen along it. I was just strolling along, trying to kick out a pace.

"I would sing songs to myself. The 'Yellow Submarine' was one of my favorites, just one of those kicking-along-through-the-forest types of songs. That type of thing- just trying to maintain my sense of humor. I would try to find something humorous throughout the day to make a joke.

"Wayne was very much a pessimist, and he didn't really appreciate my effort. I think my [optimism] had a lot to do with my survival. And I carry that lesson to this day. Wayne was very negative about a lot of things, and I was very positive. He was not very receptive to any kind of joke.

"That trip was tough. I can't think about it day in and day out. Guys were sick. Guys would fall out. Guys would be just left by the wayside. When I say seven guys died plus Wayne, some of those guys were assumptions. I mean the guy would just collapse along the trail.

"Sometimes the NVA would drag them off, and we would still be in earshot when we heard a rifle shot. There is really no doubt in my mind [this happened] to all of them. [The guards] may have waited an hour for us to get farther up the trail [before shooting] them so we would not know for sure they did not survive.

"That knowledge was in my mind when I was trying to get along under the condition I was in with my leg. My leg was just getting worse and worse.

"One day I finally just hit the limit. Beyond any mental control at all, I just collapsed on the trail. And, I couldn't get myself back up. At that moment, I knew well what had happened to the other guys that had just collapsed during the day's journey. And I knew well that I may have come to the end of my life right there. But no matter how much I wanted to live, and how much I wanted to get up and walk, I just couldn't. I was a physical disaster. My leg was about to fall off."

At that point, two of the friendly South Vietnamese prisoners grabbed Reeder and started to pick him up. The NVA Communist guards began abusing them immediately, but they persisted.

One of them was an A-10 pilot. He was short like most Vietnamese. He was in bad shape, but he pulled Reeder onto his back. He put Reeder's arms around his neck, and dragging his feet on the ground, carried him the rest of the day till they made it to the next way station along the trail. At that point, Wayne Finch was still alive.

The morning after Reeder had been carried to the way station, he got up with more pain than he had ever experienced. He was still on his feet when they tried to cross another log bridge out of the camp. He was having terrible problems with his leg. It wouldn't support him.

He lost his balance and off he went into the rushing river. Wayne jumped in and dragged him to shore, where the guards had a little conference.

For reasons we don't understand fully, the guards would dispose of a POW en route, but they didn't seem to mind leaving a POW at a way station. They made a decision. Since they were still in the way station place, Reeder would stay there.

Both Americans asked that Finch be allowed to stay and care for Reeder. Wayne was quite worried about Bill's chances for survival. All the physical problems were against him, and generally, neither of them knew of anyone left behind who survived. But the guards made Wayne go on with the group. Reeder was left alone with a single guard.

Now his luck began to change, but he did not have any idea it was for the better. For the first time since becoming a POW, Reeder got some medical care. A medic came and looked at his leg; speaking through the guard as a very poor interpreter, the medic told Reeder they were going to give him a drug. There was considerable confusion establishing the drug as penicillin.

Penicillin might be a miracle drug, but Captain Reeder had experienced a violent reaction to it a few years earlier. He had been told he was so allergic that he must never use it again, or it might kill him immediately. He explained this to the medic, and they went away and consulted some more. They came back and told him that the leg was so bad that they must amputate. Reeder was at the end of his rope.

"I just thought in my mind,'Now what in the Hell are my chances to survive if they have to take this goddamn thing off here? I am in awful shape. I am almost dead. My leg is very infected. I mean, I have a whole bunch of options and none of them are any good.'

"So I told them then, "Okay. Go ahead and try the penicillin."

So, they gave him an injection. Nothing bad happened.

The injections continued for five days; massive injections of penicillin, a couple of times a day. It should have killed him, but he experienced only minimal reaction to the penicillin. Nothing more serious than itchy hands.

Years later, after his release, doctors queried him about this curious event. The shots were *very* painful, he told them. That was their only clue.

They told him that third world countries often get uncut penicillin- straight penicillin with no buffers. A shot of penicillin without buffers is like getting a shot of glue; very painful. In the USA, people are often allergic to the buffer used to make the shot more bearable. They concluded that this had been Reeder's problem.

Anyway, after five days, Reeder was able to put weight on his ankle again, and he was once again on his way to Hanoi. The first several days of this trip were made with only him and his guard, who would march a few feet behind with his AK-47. The guard wanted to catch up with the other group, but they never did.

They found another man from their group left behind at another way station up the trail, a Vietnamese soldier. When they got there, he was very nearly dead. They stayed there one night and put Reeder in the cage with him. He could do nothing for himself; Reeder did it all.

"This guy weighed as much as a parakeet, but I still couldn't really help much. Hell, it was like throwing a blanket over your back. I hauled him out to the latrine. I sat him down, and he crapped all over himself. He died next morning. And we went on with our trip."

Eventually they joined another prisoner group, one made up of high-ranking South Vietnamese (friendly to Americans) officers. They had been captured with the 22nd Division Head-quarters when it was overrun. They had a full Colonel and the whole Division Staff, as well as a bunch of Lieutenant Colonels. The guards were also a very VIP group. The treat-ment with that group wasn't too bad, unlike Reeder's guard who never stopped abusing him.

The group walked all the way up to the North Vietnam border.

Once inside North Vietnam, the POWs moved in trucks until they became subject to more air strikes. Then they would march north by some more circuitous route. Reeder was convinced riding in trucks would be much easier. He had even been praying that once they were off the trail that they would be able to ride in trucks. He was in for a shock.

"The damn trucks had no springs, no nothing. The pain was almost as bad on my broken back. We were bouncing around in the back of those deuce and a halfs [two and a half ton Army workhorse trucks] over all these roads with bombed out craters, rocks, and everything. That was no fun, either."

They were riding in the trucks one night in an area where there were bombed out hulks of trucks on either side of the road. A flare bird popped his flares, and all those F-4s came rolling in.

"Shit, I knew that we were going to get bombed then. The trucks stopped immediately and pulled off the side of the road a little bit. They had twigs and crap sticking out trying to make them look like bushes. But then the air strike started."

Excitement lit up his face as Reeder spoke of this event.

"The goddamn truck in front of us got blown up. Then the truck behind us got blown up. You can't come a whole lot closer than that. They didn't let us get out of the truck or anything.

We just sat there while the guard held his rifle in his hands. I guess they thought we'd just run away, and probably we would have. I could just see myself getting killed by friendly air."

But he survived.

Again.

The trucks that had not been hit regrouped and continued the trip.

Then one day when they were walking through a badly bombed out area, the guards started yelling, "Bomhigh! Bomhigh!"

I was just wondering, '*What the hell are they yelling about?*' when one of the Vietnamese Officers told me that "bomhigh" means B-52 strike. The guys jumped into one of these craters and dragged us in there with them just as the B-52s came overhead.

"So, we got to the bottom of a crater. This time, we were not *close* to a B-52 strike. We were in the *middle* of a B-52 strike. That goddamn thing went off all around us! Bombs were going off to the back, to the front, to the side. All the while we were all in this crater with the guards.

"One of the officer guards pulled his pistol out and was holding it up to the head of the guy next to me. They just did not want any of us to try to escape. I don't think any of us would have tried to go anywhere.

"I had always heard before that a B-52 strike was so massive that, if it didn't kill you, you would have blood coming out of your eyes and ears. The concussion alone would destroy you."

The reality of it wasn't actually that grim. The bomb didn't hit Reeder's group or the crater they were in. They all survived, with the only aftermath Reeder remembers being his ears were ringing.

The general population was so propagandized by this time

of the war that they really hated Americans. They would attack prisoners with sticks or hoes. There were even instances when some Vietnamese captors encouraged locals to attack POWs.

It became a real chore for the North Vietnamese Army (NVA) to keep Reeder's presence hidden as they traveled through the public.

As they moved through any population at all, the guards put the American in the middle. They put the South Vietnamese prisoners around him in a circle, and the guards would be outside the circle.

Aggressive people would sometimes get through the guards.

Reeder said, "My poor South Vietnamese friends would get battered with hoes and sticks and stuff. But it kept them from getting to me except for the stray blow or rock or whatever."

The group stopped at various places along the route; maybe a little civilian village or a regular military encampment.

"In one military encampment where they stopped, the Communist camp commander got up to give this big political damn speech, and hell it went on for about 45 minutes or an hour.

"At the end of the speech, he said some other words, and one of my Vietnamese companions turned to me and asked, 'Do you know an American named Feen?' I told him no. He said, 'Oh, okay. 'Cause he died in this camp a week ago.'

"We went from there to the area where we were to be held for the night. There I found a South Vietnamese lieutenant who had been in my original group of prisoners. He came up to me, real sad. He said that Wayne Finch had died in that camp. So Feen was actually Finch, a pronunciation error.

"The lieutenant related the whole story. They had left him behind to care for Wayne. He said Wayne had always worried that I had died. Every day [he spoke of me]. He didn't see how I could make it.

"He continued to be fairly strong along the journey, but then he got malaria. And in one week, it just got worse and worse. He was crapping all over himself, and just really falling apart. And the lieutenant (Huoung) cared for him during all of this.

"They had buried him in the camp, wrapped him in a sheet of some kind when they stuck him in the ground. He died asking for a Bible. On the day he died, they offered him a bunch of food, but it was too late."

Captain Reeder's group stayed in that camp for about five days awaiting trucking transportation. He was terribly depressed over learning of Wayne's death, and his South Vietnamese friends could see that. They tried to help him out as much as they could with some of their food rations.

In fact, one of the captive Vietnamese lieutenant colonels had hidden a ball point pen inside the lining of his fatigues somewhere. He was originally from North Vietnam before coming south, and he intended to use this booty somewhere in Hanoi to barter a favor. Instead, he got the pen out and gave it to one of the guards he had befriended in trade for potatoes. This officer traded his sole valuable possession to give Reeder a little extra food. And later when Reeder was really sick from malaria, the Vietnamese officer gave Reeder almost all of his rations. They became good friends.

They got to Hanoi very shortly thereafter. It was the tenth of October.

WHEN REEDER COMMENTED on the dates, it was peculiar the way he recited them in such a neat and orderly fashion.

"We finally got to Hanoi on... let's see, I'll try to give you the dates. Well, on the second of July 1972 we left the camp in the jungle. On the Fourth of July we were two days out of that Cambodian camp and in really bad shape. When Wayne and I got up in the morning and looked at each other, he said, "Happy Fourth of July." He stuck out his hand, and that was our Fourth of July celebration. Then on the tenth of October, I arrived in the outskirts of Hanoi."

Reeder made a point of how important the date and weekday were to him. When he was first in the Cambodian POW camp, he had no idea what day it was, what day of the week or anything else.

In fact, the trauma of his shoot-down had caused him to lose a day. Until he was released, he believed that he had been shot down on the eighth of May. In fact, he was shot down on the ninth of May. When he met Wayne Finch in that cage one of the first things he told Reeder was, "Today is [some time in June 1972.] Today is Saturday."

Reeder had been really distressed over this forgotten sense of time, as though part of his identity were tied to it. As a prisoner, part of his daily routine consisted of a conversation with himself including acknowledging the day and date.

From the time he met Wayne, every day he knew what date it was and what day of the week it was. Every day, he recited this to himself so he would never again forget.

REEDER WAS STILL with this group of South Vietnamese officers, many his good friends. He was especially close to LTC Niem Kay, an engineer officer on the staff.

When they approached Hanoi, they stopped that night on a little farmstead on the edge of town. Very early in the morning, just at dawn, a jeep pulled up outside, and the Communists talked among themselves. Then they brought Reeder out, and they brought Kay to interpret.

"Here we were, standing on a low hill on the outskirts of Hanoi. We were on a farmstead with the sun rising, green countryside all around us. With a view down in this valley, the City of Hanoi. Had it been any other situation, it would have been a beautiful, beautiful sight. A storybook kind of sight.

"Kay is telling me, 'You must go now. They are going to take you to a camp where there are other Americans. We will be taken to a camp for Vietnamese prisoners somewhere else. I will never forget the friendship that has existed between my country and yours. I will never forget the friendship that exists between you and me. Goodbye my friend.'

"He shook my hand, and we parted. It was really an emotional moment. Then they put me in this jeep, in the back, on the floor, blindfolded, tied up, and off we went."

8

FINALLY: PRISON – PLANTATION GARDENS

Hỏa Lò Prison was a prison in Hanoi originally used by the French colonists in Indochina for political prisoners, and later by North Vietnam.

During this later period, it was known to American POWs as the "**Hanoi Hilton**". Wikipedia Hanoi Hilton[*]

Plantation Gardens

REEDER'S WISH CAME TRUE; he was finally delivered to a regular POW camp.

---

[*] Following Operation Homecoming, the prison was used to incarcerate Vietnamese dissidents and other political prisoners, including the poet Nguyễn Chí Thiện. The prison was demolished during the 1990s, although its gatehouse remains a museum. https://en.wikipedia.org/wiki/H%E1%BB%8Fa_L%C3%B2_Prison

"The first thing they did was take my flight suit away from me, which was in terrible shape.

"Then, they took my most prized possession- my rag. It had been my T-shirt. My OD [olive drab green] T-shirt. You have to remember, a rag was a prized possession. It allowed you to maintain whatever dignity you could muster by cleaning yourself up just a bit better than otherwise possible. I had my T-shirt as my prize rag. They took that, even though I wanted to keep it.

"I had no underwear to lose. I had surrendered it long ago. Before I got dysentery, Wayne got dysentery. When were in that first cage in the jungle, I gave him my underwear. I gave him my underwear to clean up all the crap and everything, and so my underwear became his rag.

"Then they took me over to this thing almost like a horse trough. They threw buckets of water on me and gave me a bar of lye-kind-of-soap. I washed up and took a cold-water bath- very cold.

"They sat me down on a stool out in the courtyard. A guy came up with scissors and cut off all my beard. I have a very heavy beard, and I had not shaved in five months. Then they gave me a razor, like a safety razor and some soap, and I shaved. Cut myself to shreds."

He was issued a formal prison uniform. They were the blue-gray stripped pajamas seen in many prisoner photos, with a pair of Ho Chi Minh sandals. So, he finally had something for his feet.

"Then they threw me in this solitary confinement room, all by myself. I lay in there for about an hour. It was getting on into evening time, and they brought me a bowl with bean sprouts. All I had had was rice balls. In fact, floating in the bean sprouts was a couple pieces of fatback- with skin with hair. Something was animal flesh, and just the fat. There was no meat really. And a little loaf of French-type bread.

"They had a little speaker up in one corner of the room. I found out later it was a propaganda broadcast more than anything else. But at that time over the speaker they played a song by the Carpenters, 'We've Only Just Begun.'

"I sat there and had a total, complete emotional decomposition. As I said earlier, I had set this goal, 'Get to Hanoi, and you will survive.' Well, I got to Hanoi. I was clean. I had on prison pajamas. I was sitting on the floor of this solitary confinement cell.

"But I had food! I had this music! The tears just rolled down my cheeks. For the first time since the day I was shot down, there was no doubt in my mind that I would survive. I knew that I was going home, and I had no doubt."

After the chow and the music came the propaganda broadcast. Every day in regular intervals, at least two or three times a day, the prisoners would hear various propaganda broadcasts. Some Hanoi Hanna, some Vietnamese, and there were even some Americans speaking.

The next day when they called Reeder in for his first interrogation session, they started out being nice.

Reeder grinned some as he imitated the interrogator.

"Hi, how are ya doing? What are your problems? Say, we have this camp radio. By the way, did you know we have a group of people who as an extracurricular activity here manage the camp radio and all? It gives you a good activity. It will keep you busy while you're a prisoner here. We'll let you do that."

Reeder laughed, thinking of the moment.

"But the interrogation, though never as brutal as in the jungle, quickly turned from this nice guy thing to then saying, 'Oh, okay, well, if you don't want to do that, how about just filling out this biographical form for us as part of our in-processing here at the camp.'

"And if you still didn't buy into that, and when you didn't

cooperate with any of that stuff, there came a point when they got downright belligerent."

"The Allied forces bombs started the very next day, on the eleventh of October. I could hear the bomb strikes and the anti-aircraft weapons shoot. Some of them would hit so close we literally had shrapnel hitting the compound.

"We were all very happy. The bomb strikes increased our morale. At the same time, we would [lie] there at night, cold and shaking, wondering if something would hit the building. I didn't know it at the time but they knew exactly where that camp was. They knew where most of them were."

Reeder was kept in solitary confinement for only a couple of days. The first night he heard muffled voices through the walls. He couldn't tell if they were American or Vietnamese or guards or what. He was very leery to shout out or make any communications with people.

The first night, his first recognizable tap on the wall came: "Shave and a haircut,'" and he tapped back the "two bits." He didn't recognize anything other than that.

He didn't learn until later that the cell next door was a cell of very seriously sick and wounded prisoners, probably a half dozen or so. They were trying to communicate with him using the tap code system that was known throughout the prison system. But at that point Reeder didn't know anything about the tap code. He wasn't going to learn it until later when he was with other POWs.

"During those days in solitary confinement, I was taken out each day for what turned into more of an indoctrination session. They were trying to get some propaganda type information from me, which I successfully avoided completely.

"However, they were also trying to get some biographical information from me. In a relatively short period of time, I finally did break down and fill out some biographical information for them.

"I am not proud of it, but not unproud either. I maintained well enough, I think, as a soldier in my captivity.

"I think because they were not as brutal in their interrogation as they were in the south, that it was not as easy to resist in the north. I found it very easy to resist in the south. The harder they would deal with me, the harder I would deal with them.

"I still didn't ever collaborate with the northern interrogators. I never answered any military questions. But I did, after several days of solitary confinement, fill out the goddamn bio sheets so they would get off my back and leave me alone. So that is the one thing I am not real proud of."

Reeder went on to say that a more sophisticated enemy might have used the bio sheets to harass his family. Luckily, he never experienced any of that. He was honest in his bio sheet, though he does not now recommend that. Reeder made no excuses, simply explaining that his mental and physical condition were at the outside limits of his ability to endure.

Even after he filled out the biographical paper, he still never did get to write a letter home. He never received a letter. He never got a Red Cross package.

"So, I did fill that bio data there. And at that point in the war, it got me out of solitary confinement. Had it been earlier in the war, it was routine that people spent much longer in solitary confinement.

"When I got there, it being October, negotiations were under way. I think even the Communists realized that the war might be coming to a close. So I benefited in my treatment from my captors in Hanoi because of the lateness of the war."

Part of the Plantation Gardens was a big, fancy, two-to-three story colonial home with gardens all around it. That colonial structure housed the administrative offices of the Communist camp authority.

A short distance from that structure there was a rectangle

of single-story warehouse-type buildings. These had been converted into prison cells, and that is where they kept the prisoners.

Reeder heard of no escapes out of that camp.

Part of the mentality of the prisoners once they got to Hanoi was the futility of attempting escape. Even if they went through all the things required and were successful in getting out onto the street, they were still going to stick out as a six-foot tall American in the middle of downtown Hanoi, dressed in prison garb.

"Though we were able to gain some very effective communications systems in the camp, we really didn't do anything approaching what the allied forces accomplished in prison camps in World War II as far as falsified documentation, civilian type clothes, or any of that kind of apparatus.

"We were much more restricted in what we could and could not do than some of the past prison camps in WWII and Europe. Those guys were allowed to go out and garden. They had some crafts introduced into the camp, by camp authorities. Maybe they would have had access to materials to make clothing or whatever. We didn't have any of that stuff. We were lucky to be able to secure a couple of small pencils and a few shreds of paper with which to communicate.

Only a couple prisoners spoke Vietnamese. One was a Marine Corps enlisted guy.

Reeder was moved into a room, a cell that had seven other prisoners in it. It was good to be back with other Americans, to be able talk and carry on a conversation.

They were allowed to talk to each other as long as they did not do it after lights out. They would generally try to whisper before one of the guards would invariably hear them talking. Then the guard would stick a gun barrel through the door and tell them to shut up.

The cell was just big enough so they could all sleep on

something defined as a bed. Something like planks on sawhorses held the prisoners two feet or so above the ground.

"We lay there with a blanket we could pull up over us, but it got kind of chilly in the middle of the winter time. There was no heat of any kind.

"Down on one end of the room, we had two buckets that we would use for latrine. We just did it right there in the room. Every morning, a detail from one of the other rooms would come around and empty the buckets. It was usually one of the enlisted men.

"We kept a pretty dull routine in Hanoi. We were locked in our cells all day long. We did have good latrine facilities with the buckets right there in the room, so that was no problem. We had a container of warm water in the room so we had water available all day long, and we were fed twice a day. Once about mid-morning, and once in late afternoon. Finally, we got to a diet that was other than just those two balls of rice that I had had the whole time in the jungle.

"The diet varied, but very slowly. In other words, they would give us one staple for weeks on end before they changed to something else.

"We went through pumpkin soup, cabbage, kohlrabi and then bean sprouts. Those were about the four main staples. So, for this period of the year, it might be pumpkin soup. Just boiled pieces of pumpkin in some water. They would slop that out into a bowl and give it to you. You would get that in the morning, you would get it in the evening, you would get it seven days a week for weeks on end. Then they would get out of the pumpkin season, and into the cabbage season. Then you'd get cabbage boiled in water and slopped out for weeks on end.

"The food was bland, and we were not well fed. Still, it was an improvement on what I had before. We were all skinny, but in Hanoi, I felt like I put on a couple of pounds. I was just like a skeleton before I got there.

"My leg had gotten much better because of all the penicillin treatments. It was no longer cracked and oozing pus out of the whole leg. It was pretty well resolved except just the ankle wound. That got better, too, as time went on in Hanoi. I was no longer having to walk all over the place.

"Actually, in relation to everything I had been through, it was really a very comfortable existence. Not having freedom was of course the worst part of the situation.

"As far as survival, I didn't feel threatened with my survival while I was in Hanoi. Maybe I should have because I got a series of very severe malaria attacks shortly after I got there.

Reeder had gotten malaria coming up the trail. He may have had some minor malaria attacks prior to getting there, but after settling in at Hanoi, he became extremely ill.

"My first attack was in the cell with these other guys. I always got some moral support and comfort from those other prisoners."

The malaria attacks would start in the morning when he would wake up and feel '*blah*'. In about an hour, around eight or nine o'clock, lacking his usual appetite, he'd start getting cold and feeling chills. This would last up through noontime. He would just lie there with the chills, violently shaking on his plank with his blanket. The other men would wrap all the layers of blankets they could get around him while he shook and thrashed about.

Later in the afternoon, chills would subside. Then, while he lay there exhausted, a fever would start building. It would spike rapidly, soon becoming very high, though they could not measure it. He would often be delirious during these bouts.

"That first attack of malaria lasted about three days, and I was really, really in bad shape. We had a doctor in the camp, an American flight surgeon named Cushman*.

Cushman was not in the room with Reeder, but through the communication system, he directed the prisoners in the room to find any small pieces of rag, get them wet, and put them on Reeder during the high fever periods. He also started really raising Hell with the camp authorities about the need to get him some quinine, medication for malaria.

"Finally, then, at the end of third day, the camp medic came in, scoffed and walked out. But they did get me some kind of quinine pills. They allowed me to take them just a few days until the symptoms subsided then they didn't give me any more pills.

"Well, that's not enough to cure malaria. All that did was postpone the symptoms until the next attack. I was routinely having malaria attacks once every several weeks, but after the first attack, I could get quinine pills out of them. That did help."

Another burden was that the prisoners couldn't exercise.

They would let the prisoners out of those cells each day for about 10 minutes. During the 10 minutes they were out, they had to tend to their washing, hygiene and any other business they had.

On one edge of the camp was a horse-trough-type struc-ture made of concrete, full of water. There was a rubber bucket there the prisoners used to dip water out of that trough and throw it all over themselves for a sort of bath-shower. They were allowed to do that almost every day in Hanoi. So they

---

* He was the only doctor that was captured. He went through some terrible things himself. The Communists never gave him any of the tools to take care of people medically. And about all he could do in the jungle was hold people in his arms until they died. Pretty frustrating for a physician, having the knowl-edge but not the tools.

were able to keep fairly clean. And they were allowed to wash and alternate their uniforms in that water.

Starting again in December, the B-52 airstrikes resumed in Hanoi. They were hitting right across the street from the camp in some cases.

"Christmas day, I will never forget. The camp looked like we really had Christmas Spirit. Some of the aircraft making bombing runs had dropped chaff [little metal strips which look like Christmas tree icicles.] Christmas morning, there was chaff all over the camp."

9

HANOI HILTON
THE FRENCH-BUILT PRISON

United States Air Force, Public domain, via Wikimedia Commons

ON THE 27TH OF DECEMBER, during an airstrike bombing, the authorities came into our camp. The prisoners were told, "We're moving you out of here. Collect your stuff."

Reeder recalled, "Our 'stuff' was our blanket. We rolled that up and grabbed our long and our short clothes; whatever we weren't wearing. Then we went outside and jumped up onto trucks.

"We thought they were taking us out of the city because of the bombing, out into the rural countryside somewhere. They drove around and did a pattern all over the city just trying to disorient us. Then they pulled through some gates and let us out of the trucks.

"We weren't too sure where we were. Some of the guys who had been in the prison system for a long time spread the word this was the 'Hanoi Hilton'. That was on the 27th of December. We spent the rest of our prison time in the Hilton.

"The Hilton was a prison that was built by the French. If you saw the movie Papillon, you'll have an idea of what the Hanoi Hilton was like. The prison in that movie gave me goose bumps because it was so similar to the Hanoi Hilton.

"IF YOU CAN IMAGINE that prison situated directly in the middle of a city, then you would just about have the Hilton. The walls around it were about eight feet thick, I guess, and about 20 feet high. On the tops of the wall, they had embedded broken pieces of glass along the whole length of the wall. They had three strands of hot electric wire on top of that, and every so many yards they had a guard tower. It was very secure; an old-looking type of prison.

'The small solitary confinement rooms were very similar to what they showed in that movie, Papillon. It will give you a good idea of what the prison looked like.

The Hanoi Hilton group cells were different than at Plantation Gardens. There, they had small eight-man cells. In the Hilton, they were in large cell rooms with about 30 prisoners per room.

"They moved us over to a corner called the Little Vegas section. You can see maps of the camp in the book, *POW,* by Hubbell. You will see the Little Vegas section in the corner. So even after they moved us into the Hilton, which was packed full of other American prisoners, they kept us segregated over in this one corner of the camp."

10

FREEDOM
OPERATION HOMECOMING

*Former American POWs departing from Hanoi on March 28, 1973 from
Wikipedia Commons*

IN THE HANOI HILTON PRISON, rumors were spreading that
negotiations to end the war had begun and were ongoing. In

fact, Reeder's group had some late-caught prisoners* who reported some of the more recent news.

"Our hopes were high," Reeder remembers.

One night, one of the guards was walking around late at night with a portable radio playing, which he was not supposed to be doing. This one Marine (who spoke Vietnamese and could interpret) overheard the news report that said that agreements had been reached and signed between Henry Kissinger and Le Doc To. He heard that the war would be over. The withdrawal was already scheduled.

The next day, the camp authorities denied it. They said that the Marine did not speak good Vietnamese. They said the rumor was simply not true.

"He speaks like a child."

Nevertheless, the very next day, the camp commander called all the prisoners into the prison yard at the same time. He sat everyone down and had cameras poised.

"We had already spread word through the communication system that we would do our best to avoid letting them make propaganda out of anything with the release.

"He made the announcement in English that an agreement had been reached and we would all be going home soon. The cameras panned us to catch us with all those joyous chuckles and expressions. We just sat there with blank expressions on our faces. It really made him mad. First, he thought we didn't understand. He repeated it. One of the other guards repeated it. Then, he stomped off."

Regardless, the treatment of the prisoners steadily improved. They began letting prisoners out of their cells into the courtyard. They could exercise. They could walk around. The food improved.

---

* The only Navy guy in our group was captured the day the peace accords were supposed to go into effect.

"There at the end, they were giving us meat and fruit and all the bread we wanted to eat. They really tried to fatten us up, and I think it worked. We were hungry enough that we ate.

"We may have taken the line of thinking that 'Hell, we don't want to come back looking like we did in captivity.' The men were so half starved that they did eat."

Reeder put on a lot of weight in the last 30-60 days before being released.

Embarrassments of the Enemy

According to Reeder, there were two facts that would have embarrassed the North Vietnamese to the world.

"Most importantly to us was the fact that after 10 years of war they had so few prisoners left, especially US Army prisoners. In total, they had only 77 Army guys. I guess we had about 50 in our camp, and there were about another 20 that were held back in the jungles of South Vietnam."

Although the NVA had captured many more men, most did not survive. The treatment of prisoners was not conducive to survival. Many men died in the jungle camps in the south. Many Men died trying to make the journey to the north. So, they had virtually no prisoners left to show for the lengthy war.

The second embarrassment was that the captives were supposedly prisoners of the Viet Cong, yet they were being held inside the country of North Vietnam, policed by the regular North Vietnamese. That would have been a serious embarrassment.

Release of Prisoners, 1973

The prisoners were released in groups at 15-day increments. There were a total of four release groups. They were supposed to let the longest held guys go with the first groups, but they didn't. Reeder couldn't see any rhyme or reason to the order. It was apparently random how they picked which men would go when.

"I ended up in the very last group to come out. They told us ahead of time that our group was supposed to go home on the 24th of March."

On the 24th of March, the men assembled, ready to leave. The captors didn't say anything. Nothing happened all morning. The prisoners were not brought food either. They had no idea what was going on.

Then, about 2:00pm, the camp commander called the men out to the courtyard. At this point, all the other release groups had gone. He sat all the men down, and began announcing.

"There has been a problem with the negotiations. Nixon and the American warmongers have done an about face on the agreements. The war continues. You must stay prisoners forever. Go back to your cells."

The men were marched back to their cells, very disillusioned.

Reeder and the other prisoners knew about the Sontay Prison raid. Word had spread through the whole group. (The Sontay raid was a famous American raid to rescue prisoners. The Americans got in and out without many casualties, killed a bunch of enemy, and stunned the world with boldness of the move. But there were no prisoners for them to rescue there.)

After the raid, North Vietnam changed prisoner policy and added security because of the threat this raid posed to their internal security.

Reeder and the other prisoners were disappointed at not being released, but none of them believed they would stay in Hanoi and rot forever.

"If the agreements *had* in fact broken down, and the war was continued, we thought then we'd see the 101st Airborne come parachuting down into the skies of Hanoi, coming to get us out."

Though the men were disappointed, they were not completely demoralized. They simply went on about their routine, perhaps "hanging their heads a bit."

But then on the 27th of March, the prisoners were called back out again. They were told, 'Get ready to go.' They were issued civilian clothes with pants and shoes and a shirt. They were loaded onto a bus and driven out of the Hilton.

They were intentionally driven by some of the most severe bomb damage in Hanoi on the way to the airfield. Then they were taken to the airport there in Hanoi. There they were returned to American control in a very ceremonious fashion, one at a time.

They were then escorted out to a C-141.

"It was a beautiful sight. There, sitting in the capitol of a Communist country. a big silver C-141, just parked there. Then, we were all on board and took off. Then, we were airborne, a little ways off the coast of the Philippines.

"Finally, the pilot made an announcement over the intercom that we were outside SAM [surface to air missile] range.

"And everybody went berserk."

11

WILLIAM S. REEDER, JR.
COLONEL O-6, U.S. ARMY

U.S. Army 1965-1995, Veteran of Cold War 1965-1991,
Veteran and Vietnam War 1968-1969, and POW 1971-1973

BILL REEDER WAS BORN in 1945 in Lake Arrowhead, California. He enlisted in the U.S. Army on August 30, 1965, and was commissioned a 2d Lt in the U.S. Army through Artillery Officer Candidate School on August 16, 1966.

His first assignment was with 6th Battalion, 21st Artillery Regiment of the 5th Mechanized Infantry Division at Fort Carson, Colorado, from September 1966 to October 1967, followed by fixed-wing aviator training on the OV-1 Mohawk from October 1967 to June 1968.

After that, Capt Reeder served as a Mohawk pilot with the 131st Aviation Company in South Vietnam from October 1968 to November 1969, and during this time he was shot down on March 1, 1969, but was rescued the same day.

He then finished his Bachelor's degree at the University of Nebraska, followed by rotary-wing (helicopter) aviator training from July to September 1971, and AH-1 Cobra training from September to December 1971.

His next assignment was as an AH-1 pilot with the 361st Aviation Company in South Vietnam from December 1971 until he was shot down on March 9, 1972, managing to evade the enemy until he was captured and taken as a Prisoner of War on March 12, 1972.

After spending 320 days in captivity, Capt Reeder was released during Operation Homecoming on March 27, 1973. He was briefly hospitalized in Colorado, and then attended the Field Artillery Officer Advanced Course at Fort Sill, Oklahoma, to May 1974.

His next assignment was as a Battery Commander Officer with the 1st Battalion, 11th Field Artillery Regiment at Fort Lewis, Washington, from September 1974 to June 1975, and then as an Aerial Surveillance Officer and staff officer with Headquarters U.S. Army Europe and with 7th Army in West Germany from November 1975 to November 1976.

He then served as Executive Officer of the 334th Aviation Company, 11th Aviation Battalion, in West Germany from November 1976 to June 1977, and then as S-3 Officer with the 11th Aviation Battalion from June to December 1977. Maj Reeder attended Armed Forces Staff College at Norfolk, Virginia, from February to June 1978, followed by service as a UV-18B Twin Otter pilot and TAC Officer at the U.S. Air Force Academy from July 1978 to July 1981.

Col Reeder's next assignment was as a TAC Officer with the Defense Mapping Agency in Washington, D.C., from August 1981 to June 1983, and then as Deputy Brigade Commander and Executive Officer with the 9th Cavalry Brigade at Fort Lewis, Washington, from July 1983 to April 1985. This Fort Lewis assignment is where the author had the privilege to become acquainted with the man himself.

He served as Commander of the 3rd Squadron, 5th Cavalry

Regiment at Fort Lewis from April to December 1985, and then served as Aviation Colonel Assignment Officer with the U.S. Army Personnel Center at the Pentagon from December 1985 to March 1987.

His next assignment was as Commander of 5th Squadron, 17th Cavalry Regiment at Fort Hood, Texas, from March to August 1987, and then as Commander of 1st Battalion, 3rd Aviation Regiment from August 1987 to September 1988.

Col Reeder served on the staff of III Corps at Fort Hood from September 1988 to January 1989, followed by Naval War College at Norfolk, Virginia, from January 1989 to August 1990.

His next assignment was as Commander of the AH-64 Apache Training Brigade at Fort Hood from August 1990 to October 1992, and then as a Brigade Commander with III Corps from October 1992 to April 1993.

Col Reeder's final assignment was as Deputy Chief of Staff, at Fort Clayton, Panama, from April 1993 until his retirement from the Army on March 1, 1995.

Silver Star Citation

For gallantry during action against North Vietnamese Army forces on 14 April 1972, while serving as a helicopter pilot supporting the combat actions of friendly Vietnamese forces in Kontum Province, Republic of Vietnam.

The outnumbered ground forces were under heavy artillery fire, surrounded by anti-aircraft weapons and under attack by enemy forces. Captain Reeder took out a number of lethal guns, all while under intense fire from multiple anti-aircraft positions and small caliber weapons.

After rearming and refueling, Captain Reeder's team voluntarily returned to once again engage the enemy. In extraordinarily poor conditions of low clouds, haze, smoke and deepening darkness, his team prevented the friendly force from being overrun.

Captain Reeder's actions contributed to the escape of dozens of friendly forces and one American soldier. His extraordinary heroism and selfless sacrifice reflect great credit upon himself, the 361st Aviation Company and the United States Army.[1]

---

1. This information was gratefully collected from Veteran Tributes at https://veterantributes.org/TributeDetail.php?recordID=1362

ABOUT THE AUTHOR

E. Daniel Kingsley CW4, US Army, Retired.

Kingsley is a family man who gave his best years in service of our country. Adored by his six grown children, dozens of grands and now multiple great-grands, he has amassed a legacy of family connection.

Kingsley has published *We Came to Dominate*; the detailed recounting of the 1988 U.S. Precision Helicopter Team who dominated the world. A must read for competitive helicopter pilots, and never was there one that wasn't competitive.

Kingsley has published the series *Moby Dad; tales of the Great White Father;* a collection of family anecdotes, drenched in wholesome humor and heavy on the sentimentality.

He now lives happily with his adoring wife, Marie, in Buffalo, New York.

in

www.ingramcontent.com/pod-product-compliance
Lightning Source LLC
Chambersburg PA
CBHW032051040426
42449CB00007B/1063